MW00723527

SCIENTISTS WHO STUDY
OCEAN LIFE

Mel Higginson

The Rourke Corporation, Inc.
Vero Beach, Florida 32964

© 1994 The Rourke Corporation, Inc.

All rights reserved. No part of this book
may be reproduced or utilized in any form
or by any means, electronic or mechanical
including photocopying, recording or by
any information storage and retrieval
system without permission in writing from
the publisher.

Edited by Sandra A. Robinson

PHOTO CREDITS
© Mel Higginson: cover, title page, pages 7, 8, 12, 13, 15, 17, 18;
courtesy of Connecticut Department of Environmental Protection:
pages 4 and 10; courtesy of Marine Biological Laboratory at
Woods Hole, page 21

ACKNOWLEDGMENTS
The author thanks Mote Marine Laboratory, Sarasota, Florida,
for its cooperation in the preparation of this book

Library of Congress Cataloging-in-Publication Data

Higginson, Mel, 1942-
    Scientists who study ocean life / by Mel Higginson.
      p. cm. — (Scientists)
    Includes index.
    ISBN 0-86593-371-5
    1. Marine biologists—Juvenile literature. 2. Marine biology—
Vocational guidance—Juvenile literature. [1. Marine biologists.
2. Marine biology—Vocational guidance. 3. Occupations.
4. Vocational guidance.] I. Title. II. Series: Higginson, Mel, 1942-
Scientists.
QH91.45.H54 1994
574.92'092—dc20   94-6999
                                    CIP
                                    AC

Printed in the USA

TABLE OF CONTENTS

Scientists Who Study Ocean Life   5
What Marine Biologists Do   6
Kinds of Marine Biologists   9
Where Marine Biologists Work   11
The Importance of Marine Biologists   14
Marine Laboratories   16
Discoveries of Marine Biologists   19
Learning to Be a Marine Biologist   20
Careers for Marine Biologists   22
Glossary   23
Index   24

SCIENTISTS WHO STUDY OCEAN LIFE

Oceans are places of great beauty and mystery. Sharks live in oceans, along with whales, sea turtles and thousands of other kinds of creatures.

Oceans are homes-away-from-home for **marine biologists.** Biologists are the scientists who study living things. Marine biologists study the plants and animals of the marine world — the oceans.

Marine biologists study all types of sea life, from whales and seals to tiny **organisms.**

Marine biologists study the life in oceans

WHAT MARINE BIOLOGISTS DO

Scientists who study ocean life find out what lives in each part of the ocean. They learn how the oceans' plants and animals live together in **communities.**

Marine biologists find out where ocean animals travel and where they raise young. They study the problems caused by pollution in the ocean. They also look at how ocean life can help human life on land.

Scientists look for ways to solve the problems that cause great numbers of fish to die suddenly

KINDS OF MARINE BIOLOGISTS

Each marine biologist has a special interest. One scientist may study an important food fish, such as the tuna. Another may study the behavior of sharks.

Marine biologists study ocean plants, like seaweed. They study the mysterious lives of sea turtles and elephant seals. Some marine biologists study organisms so small they can't be seen without a microscope.

A marine biologist walks among the elephant seals he studies on the California coast

WHERE MARINE BIOLOGISTS WORK

Life as a marine biologist usually means life near, or even floating upon, the sea. For short periods of time, marine biologists may plunge into the sea!

Where a marine biologist works depends on what the scientist wants to study. Some work from boats and tiny submarines. Others work along ocean shores.

Sooner or later, marine biologists work in **laboratories.**

Shipboard scientists prepare to study a sample of sea life

By tagging sea turtles as they crawl ashore to nest,
scientists have learned more about them

Scientists look for ways to speed up the growth of food fish that they raise in tanks

THE IMPORTANCE OF MARINE BIOLOGISTS

The life of the ocean is important to everyone. Ocean life supplies us with food and fun.

However, the marine life in many parts of the world is in trouble. Pollution and too much fishing are major problems.

Marine biologists help us know where and when the ocean is in trouble. They also look for solutions to the problems they find.

Results of laboratory work help scientists solve marine problems

MARINE LABORATORIES

Part of the search for solutions to marine problems takes place in laboratories. There, marine scientists study information from their libraries, computers and photographs.

At seashore labs, marine biologists often raise some of the plants and animals they study. Scientists can also keep and study creatures that have been caught in the wild.

Scientists at Mote Marine Laboratory in Sarasota, Florida, try to nurse a pygmy killer whale back to health after it ran aground

DISCOVERIES OF MARINE BIOLOGISTS

The mighty oceans cover mountains and canyons — and nearly three-fourths of the Earth's surface. The oceans hold many secrets. Marine biologists study ocean life to discover some of those secrets.

Much of what marine biologists learn helps us protect marine life. Marine biologists also use their knowledge to help people. For example, some of the materials found in marine life are used to produce medicine.

Discoveries of scientists help marine life and human life

LEARNING TO BE A MARINE BIOLOGIST

People who grow up along seashores often want to be marine biologists. They certainly have a head start! Anyone with an interest in ocean life can study marine biology.

Most students take their first marine biology classes in high school or college. Only a few high schools and universities offer marine biology classes.

Becoming a marine biologist usually means spending more than four years in college. Many marine biologists earn at least a master's degree before getting a job.

A student of marine life studies a squid

CAREERS FOR MARINE BIOLOGISTS

Marine biologists usually work for the government, a university or a marine laboratory.

Some marine biologists are researchers. They learn and write about a special subject. Others do research, but also teach classes about marine life.

Marine biologists with ideas for projects can write to the government for **grants.** Grants pay for doing the research.

Glossary

community (ka MYU nih tee) — a group of wild plants and animals that live together and share the same area

grant (GRANT) — a gift of money to be used for a special purpose

laboratory (LAB rah tor ee) — a place where scientists can experiment and test their ideas

marine biologist (muh REEN bi AHL uh gist) — a scientist who studies the plants and animals of the marine (ocean) world

organism (OR gan iz um) — a living thing

INDEX

animals   5, 6, 16
boats   11
computers   16
fish   9
government   22
grants   22
laboratories   11, 16, 22
libraries   16
marine biologists   5, 6, 11, 14,
    16, 19, 20, 22
marine scientists
    (see *marine biologists*)
medicine   19
microscope   9
ocean life   6, 14, 19
oceans   5, 11, 14, 19

photographs   16
plants   5, 6, 9, 16
pollution   6, 14
scientists   6
sea (see *oceans*)
seals   5
    elephant   9
seashores   20
sea turtles   5, 9
sharks   5, 9
tuna   9
universities   20, 22
whales   5